THE MAKING OF TUBBY DASH

By the same author

*Technology and the Big House in Ireland
c.1800-c.1930*

One Man's Garden Railways

THE MAKING OF TUBBY DASH

Charles Carson

ISBN-13: 978-1979878067
ISBN-10: 1979878064

Copyright © Charles Carson 2017
All rights reserved. No part of this publication may be reproduced, stored in a retrieval system or transmitted, in any form or by any means, without the prior written permission of the author, nor be otherwise circulated in any form of binding or cover other than that in which it is published and without a similar condition being imposed on the purchaser.

charlescarson.wordpress.com

To my wife Ann, the love of my life

Acknowledgements

With grateful thanks to my son, Alan, who made the Phoenix-like rebirth of an extinct garden railway possible and to my wife, Ann, who helped me make it happen.

Contents

Prologue
Goodbye Marie, Hello Tubby
Rolling Chassis
Motion Work
Cab
Saddle Tank & Smokebox
Controls
Smoke Unit
Driving Trolley
Windy Hill Railway II
Summer Days on the New Railway
Hedge-Conditioned Home Brew
The Railway Room
Ghouls and Wraiths
A Blink in Time

Abbreviations
Appendix I – VI

Prologue

"You have control", said the tinny voice in the earpieces of my leather flying helmet.

"I have control", I responded into the rubber mouthpiece, gripping the aluminium joystick between my knees a little more firmly.

The silver Tiger Moth, G-AIRR, was on approach to Newtownards Airport at six hundred feet altitude. The airstream whistling in the rigging wires as we descended at fifty-five knots with the Gypsy engine throttled right back.

Coming in over the town towards Runway 22 the placid waters of Strangford lough lay over to our left. Just as the blades of grass at the side of the runway were distinguishable, rather than a green blur, I pulled back on the stick to flare out the glide. For a few seconds the aircraft's nose was pointing at the blue sky, then there was a gentle bump as the wheels and the tail skid contacted the grass strip beside the runway in a lucky text book three-point landing. The trundling of the wheels stopped. There was the smell of new mown grass and in the distance above the noise of the idling engine a far-off tractor could be heard.

The front cockpit flaps were thrown open and a leather-helmeted head appeared with a dangling Gosport tube above an old leather flying jacket. As the figure passed my rear cockpit it leaned towards me.

"Do two circuits and bumps. She'll climb much faster without my weight. You'll be fine." He said.

Descending from the wing-walkway the small rotund figure of 'Tubby' Dash, the Chief Flying Instructor of Newtownards Flying Club stumped away towards the distant hangars.[i]

So this is it I thought, as my left hand opened the throttle in preparation for take off on my first solo flight.

Goodbye Marie, Hello Tubby

The 7¼" gauge steam locomotive Marie Estelle, completed after many years of workshop effort, had proved to be a disappointment. When a few children arrived requesting a ride on Grandpa's train it was quite a chore to get a fire going and sufficient steam pressure up whilst they clamoured to get aboard. It was also difficult during a trip to manage the fire and boiler water-level and at the same time try to keep young arms and legs inside the coach. Some poseurs attempted, and succeeded, in standing upright arms akimbo at the rear of the longitudinal seat.

    Well, all that was far removed from my dreams of gentle chuffing amongst the garden shrubbery pulling a train of cherubic children. The lovely Marie Estelle had to go. She was palleted and shipped off to be sold in England early in August 2015.

    It was during the dark days of January 2016 that the idea was mooted to build a steam outline Porter locomotive as a replacement. It was to be similar in size and style; with the same 0-4-0 wheel arrangement, 'diamond' chimney stack and a wooden cab. The 0-4-0 layout with two axles and four wheels that are driven, in this case by coupling rods, produces the simplest chassis.

Plate 1. Envelope sketch

A rough sketch on the back of the proverbial envelope covered the essential features and proportions. It was to be powerful enough to tackle gradients with several passengers entrained but small enough to travel in the back of a hatch-back car. Adhesive weight would be taken care of by the combined weight of the 24V traction and the 12V 'ancillaries' batteries to be situated under the saddle tank.

The ancillaries were to be numerous. This electric locomotive was intended to be as good as a steam one if possible. There was going to be a large headlight as fitted to this type of narrow gauge locomotive, cab lights, flickering firebox lights and synchronised audio chuffing and smoke effects.
To visually replicate the steam engine driving motion the locomotive was to be fitted with dummy steam cylinders with moving piston and connecting rods. All this was to try to achieve as authentic an experience as possible. The colour was to be matt black and a pair of 'Stars and stripes' flags were to be mounted on the front beam as befitted a locomotive of the Porter

Locomotive Company of the USA.

What about a name? After much consideration, Tubby Dash was decided upon. 'Tubby' as a reflection of the portly shape due to the short length and saddle tank, and 'Dash' for the hoped for turn of speed when necessary. Hopefully the real Tubby Dash would have approved.

A "Google' search of the Internet provided several useful ideas regarding constructional methods and also how effective a 'chuff' sound-card with a powerful audio amplifier could be. It was helpful to see that one gentleman in sunny California had built a Porter locomotive using similar techniques and materials to those I intended to use; namely ¾" plywood, sheet steel and a MIG welded chassis.

A four-wheel driving truck was also to be constructed that would be comfortable for myself as a six foot driver with two knee replacements. It was also to have reliable brakes. Like the locomotive it was to be finished in matt black. No gaudy colours here; panache was what we were aiming for, as in top hat and tails!

The whole assemblage was to be completed in about three months and built inexpensively using available materials where possible. I have always been impressed by a quote in the front matter of the Nevil Shute novel *Trustee from the Toolroom*. 'An engineer is a man who can do for five bob what any bloody fool can do for a quid'. A definition of unknown origin that in new money would read twenty-five pence and one pound respectively. A great pleasures in Model Engineering or one of its ubiquitous cousins, DIY, is making appropriate use of 'found' items or recycling humble materials and putting them to new and more exotic uses.

The project was nearing completion when a family decision was taken to hold a midsummer party and there 'had' to be a railway for the many children expected to attend. Well, there was no railway due to circumstances described in *One Man's Garden Railways.* Could one be built within the remaining three months? A subsequent chapter details the efforts to meet the deadline. First however, the construction of Tubby Dash.

Rolling Chassis

I feel that there is a natural sequence of events for building miniature locomotives and the first step is to make the frames so that the wheels, axles and axle boxes can be assembled with the goal of achieving a rolling chassis.

In the dim and distant past, around fifty years ago, I could face cutting miniature locomotive frames from ¼" thick steel plate using a hacksaw, chain drilling the outline of the axle box slots and removing material with a cold chisel and then finishing with a file. Nowadays however, at least one different method of construction in the home workshop is possible with the acquisition of a small MIG welder.

So, off to the local steel suppliers to purchase a six meter length of 20mm X 6mm flat black steel. For a small charge they cut it into three lengths to go into the car. My mitre band-saw made short work of cutting up the pieces required for the 30" long chassis. Although care was taken to make a ¾" plywood jig to hold the parts during welding, movement took place of the vertical horn pieces and the resultant slots were a hopeless fit for the sliding axle boxes. There was nothing for it but to attack the welded up chassis frames with a 4½" angle grinder. Fortunately, by using spacers and multiple clamps success was achieved at the second attempt. See Plate 2.

The ground level track in the background is the Rocklands Railway now lifted and relocated to a new venue as described in a later chapter.

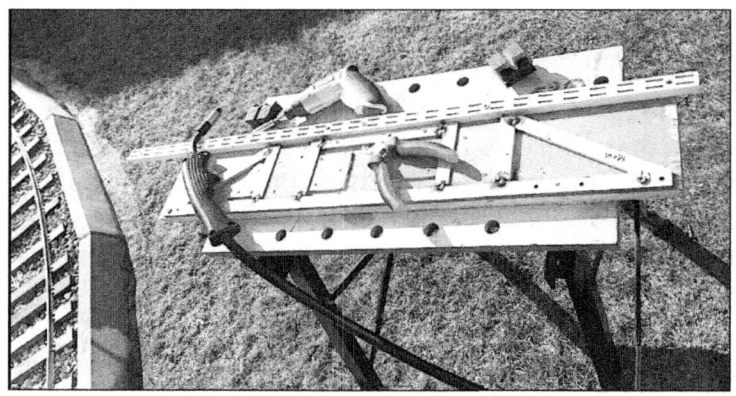

Plate 2. MIG welding frames

Next the frame steelwork was spray painted with zinc rich primer. See Plate 3.

Plate 3. Painted frames

Axles next. These were turned from 1⅛" round bar

that was in stock even though it was rather big in diameter. The extra material would always add to the locomotive adhesive weight. Unfortunately there was no suitable stock for the wheels. A visit to a reclamation yard deep in County Down produced a length of 5" diameter round bar sufficient for four 1⅛" wide narrow gauge wheels. It, along with rather rusty pieces of ⅛" and ¼" steel sheet cost £10.00. As on previous occasions the bar was sliced up using the band-saw and turned to suit the axles in the same way as described in *OMGR*, pages 81-84. Crank pin centres were marked out whilst the wheels were mounted in the lathe to ensure accuracy and subsequently drilled and reamed $^5/_{16}$" diameter.

Plate 4. Wheels and axle boxes

As a 'guesstimate' it was decided to employ eight ¼"

ID X 16 SWG coil spring for suspension. Inspired by the sunny Californian locomotive builder, 'Axle boxes' were made up from square pieces of ¼" plate that were recessed for flanged ball-races and sized to be an easy sliding fit between the horn guides. Screwed on 'wings' of 18mm X 3mm flat steel retained the bearings in the axle boxes and the axle boxes in the frames. See Plate 4.

Short stubs of ¼" rod in drilled holes in the top edges located the bottom of the springs. The top of the springs were held in place by the heads of 4 BA cheese-head screws fitted in holes drilled and tapped in the frame member above. Axle box keeps, ⅜" X ⅛" steel strips were screwed to drilled and tapped holes in the bottom of the frames with 4BA Allen head screws

The basic chassis was completed with eight ⅜" X ¾" X ⅛" angle pieces at the ends of the frames secured with 5mm bolts and screwed to the 1¾" X 7/8" keruing wooden beams with 4mm X 20mm C/S wood screws. Spacers and the motor mount were constructed with 20mm X 4mm flat and ⅛" angle steel. The slots in the motor mounts were chain drilled and filed out. See Plate 9.

At this stage an electric motor was selected. See Plate 5. From previous experience it had been found that electric scooter motors and controls were suitable for locomotive traction and accordingly a search was made on eBay. This time however, it was decided to employ a ⅓ HP 24v geared motor to simplify construction by removing the need for a lay shaft. However, to obtain a reasonable track speed, say a maximum of six miles per hour (possible need to 'dash'), the originally fitted small sprocket would be unsuitable as calculation indicated that a near 1 : 1 ratio was needed.

Plate 5. Traction motor

Plate 6. Sprocket and hub

A collar was bored 1 ⅛" to fit an axle with a 4BA grub

screw fixing and four 6BA threaded holes on the face. This was to allow the later fitting of an 'impulse switch' for synchronised 'chuffing'. The axle ends were drilled and tapped 6BA for connecting rod retaining discs. These were made from 'mudguard' washers with a C/S hole. With a sprocket fitted to one axle (the other axle to be driven via connecting rods) and the collar in place the wheels were fitted to their axles using metal adhesive ('Pacer' ANL-RC/HT). It is essential to 'quarter' the wheels so that they will turn without tight spots with the coupling rods mounted. That is, by convention, the crankpins on the right hand side leading by 90 degrees to those on the left.

Plate 7. Quartering jig

The simplest way to achieve this (nerve racking) task is to use a plywood jig in the lathe when 'glueing' the wheels to their axles. See Plate 7. The bottom strip fits between the lathe bed-ways and holds the jig central. Each axle is mounted between centres and the wheels rotated such that the crankpins are against the opposite angled edges of the 'V' cut out.

It is worth rehearsing the actions several times before applying the adhesive. This is to avoid having to close operations in the workshop for several days whilst solace is sought via copious home-brewed beer consumption. Fortunately, following my own advice, a happy result was achieved.

After a fair amount of work, the collection of bits when assembled becomes a rolling chassis that can be pushed up and down a length of track. Luckily, there were no white-coated observers to pronounce on the sanity of a seventy-something man (boy?) 'chuffing' his heart out in the workshop.

Motion Work

One of the many attractive features of a steam locomotive is the motion work. It exhibits muscular and animalistic movement as the connecting and coupling rods perform their oscillatory functions of transferring the steam power in the cylinders to the wheels. To be a 'proper' steam lookalike engine, Tubby Dash had to have this desirable feature. The provision of the crankpin holes in the wheels to enable this have been described in the previous chapter.

As he is an electric locomotive, the piston and connecting rods are for appearance only, doing no work. The coupling rods however, do transfer power from the rear wheel-set to the front one. The rods were made from 20mm X 6mm flat steel with bored out slices of 1⅛" diameter bar welded to the ends for bearing housings. To minimise power loss due to friction both sets of rods are fitted with ballraces. These were obtained inexpensively as skateboard bearings on eBay. The 'small ends' driving the piston rods via the crosshead, guided on ½" square steel slide bars, are fitted with phosphor-bronze bushes.

The dummy steam cylinders are pieces of 2¾" OD round plastic rainwater down-pipe with turned birch plywood cylinder covers. Brass 'glands' are fitted to these for the 8mm diameter stainless-steel piston rods to slide in. Cleaned up pieces of the ⅛" steel sheet, ex-reclamation yard, were used to make the crosshead sides. The ½" X ¼" spacer/slippers were sawn from phosphor bronze pieces from the scrap box. They were then milled in the lathe to a few 'thou' wider than ½" for a sliding fit on the slide bars. Eight 6BA screws were used to assemble each crosshead once the connecting rod was in place on its pin. See Plate 8.

Plate 8. Motion work

The complete assembly could now be rolled up and down a length of track to check for any serious binding or tight spots. There was seen to be reluctance for the wheels to turn with some slipping on the track. Probably this is due to lack of adhesion weight as yet and the need for 'running in'.

Plate 9. Chassis motor mount

By temporarily fitting the motor complete with its drive chain and a battery balanced on the frames connected to the motor via the control box, we now have a (rudimentary) locomotive to try out on the track and quietly make a few more chuffing noises. See Plates 9 and 10.

Plate 10. Motorised chassis

Cab

Now with the 'works' completed it is time to move on to the upper structure of the locomotive. The cab of most Porter narrow gauge engines was of wooden construction. This is an opportunity to try out one's joinery skills. However, there should be no difficulty if an inexpensive compound mitre-saw is available.

The cab floor, 12" wide X 13" long, was cut from ¾" birch plywood as some offcuts were available from a previous project. This superior, but more expensive, material cuts cleanly and has no missing bits in the layers. Using the mitre-saw, ¾" ply pieces were cut that when glued together formed the cab front with the long windows. This avoided tedious edge sanding that would have been necessary if the windows had been jig sawed out of one piece of ply. The 15" high sides were formed from 2" X 1" PAR lath screwed together through counterbored holes. Where the side panels are fitted in a 10" X 4½" opening, the lath was ⅛" grooved on the edges to frame the 3mm ply panels. Grooving was accomplished on the circular saw table with the ripping guide set to half the timber thickness and a saw depth of ⅜". The panels themselves were made to resemble planking by carefully depth setting the mitre saw to just mark the face.

Covering the cab is the 16¼" long X 14⅛" wide curved roof made from 20 SWG steel. These dimensions allow for a ¼" overhang along the sides and front. This material was purchased from the local hardware merchant as a 2m X 1m sheet. It was brought home in/on a small car trailer by 'bowing' it until is sat inside the trailer sides *a* la gypsy caravan. This method of transport succeeded in that the bending of the sheet was within the elastic limit of the material and it more or less resumed its normal flatness when released from the trailer. I was also thankful that due to being well roped down it also did not take flight during transit at thirty miles an hour.

A beam compass was made up from a wooden lath by drilling a hole at one end a tight push fit for a pencil and at the other for a nail at the desired radius for the compass point, in this case about 32". Two curved roof members were marked out on ¾" thick pine and sawn to shape with a jig saw. Finishing the outer curve was accomplished using my vertical sander. The concave or inner curve was successfully tackled using a 4" hole saw in the vertical drill with aluminium-oxide paper wrapped around the drum and attached with contact adhesive. The steel sheet was hole punched around the edges and drilled in the unreachable areas ⅛" diameter. All 44 holes were countersunk to suit 3mm x 12mm crosshead woods screws. By starting in the middle of the roof and working outwards the curve was pulled without having to roll the metal. The roof was spray painted with a grey primer and then finished matt black. All woodwork was given two coats of ebony wood stain. Finally the cab was attached to its base by steel angle brackets. Small steel parts, such as the brackets and zinc-plated screws, were 'blacked' by a short immersion in a bath of "Haematite' cold patinating fluid for steel and iron.

An important part of the romance of railway locomotives lies in the name. Whatever the appellation proudly displayed on polished brass or other plates alludes to, it creates an expectation. I am sure we can all think of examples that evoke a mystical world of travel to exotic or just well loved places by rail.

How to name our own locomotive? Well, it is hoped that the name reflects the thought processes that went into the origins of the idea to construct it and perhaps also hopes for its performance.

A present day recollection of an event that occurred half a century ago was germane to this name; Tubby Dash. It is recounted in the Prologue.

Several of my previous locomotives carried etched brass name plates. These were made in the workshop using 'Letraset' peel-off letters stuck onto brass sheet and immersed in a Ferric Chloride etching solution. Although the finished product with the raised polished letters, is prototypical and pleasing to the eye, it is a tedious process. Also the fumes from the reaction can cause rusting of any steel tools or bits-and-pieces nearby, if the process is carried out in a poorly ventilated area. Another problem is the hazardous Ferric Chloride resulting from the process. which needs to be disposed of in a proper manner and not just poured down the drain. Since the home computer is pressed into service for many tasks these days, it seemed appropriate to give my Mac a go at producing locomotive nameplates.

The name plates were to be 6½" X 1⅝" and have letters about ¾" high. It was found that the 'Poor Richard' font of 70pt size with black lettering on a gold background produced the most pleasing result. This font seemed to reflect the tubby nature of the locomotive. Print out was on 220gsm white card using my inkjet printer.

The card 'plates' were guillotined to size and stuck to 3mm plywood backing and given several coats of weather proofing varnish. The plates were mounted in the middle of each of the 'planked' sides by means of No.4 X ⅜" RH brass screws with the protruding points sanded off inside the cab. See Plate 11.

Plate 11. Nameplate

Saddle Tank & Smokebox

The saddle tank and smokebox were constructed as one unit that would lift off together to give access to the batteries. There are two 12V X 12Ah batteries in series to supply the 24V traction motor and one 12V X 12Ah one to supply all the 12V ancillaries. All the sheet steel work is 20SWG and where necessary rolled in a set of 12" geared rolls. Construction of the saddle tank began with the curved top part which is of 4" radius screwed to a front ply bulkhead by 3mm X 16mm CSK wood screws and raised up on 3⅜" vertical sides. See Plate 12.

    A ply arch supports the rear of the tank and sits against the cab front. On top is 3" diameter X 4" high steam dome, closed with a birch ply top. The smokebox was rolled as a 6" diameter X 4½" long cylinder. Six angle brackets at the front attach the front door-ring and six to the rear screw it to the ¾" ply bulkhead at the front of the saddle tank.

    Porter locomotives usually had a tall chimney with a 'diamond' stack and a large headlamp sitting on an ornamental bracket attached to the front of the smokebox. Dealing with the chimney first. It was rolled up to 2" diameter X 8" long and spot welded. A supporting saddle was formed by bending a square piece of ⅛" sheet steel over a hardwood former in the vice.

Plate 12. Sheet metal work

The saddle was then mounted on a wooden faceplate by its four mounting holes and bored to admit a short collar that was a sliding fit for the bottom of the chimney. An interesting item to make was the diamond stack. Plate 13 shows the development of the top and bottom cones.

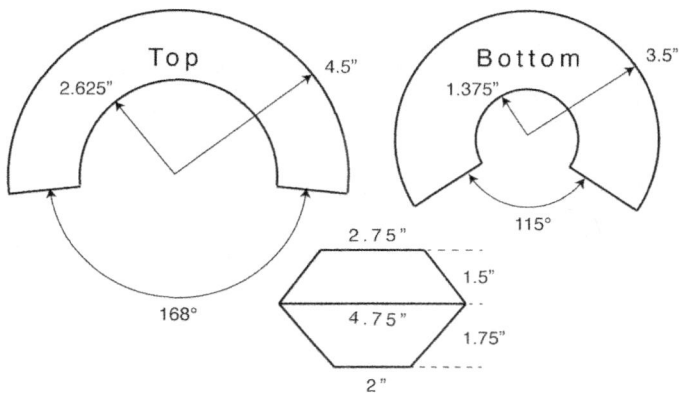

Plate 13. Diamond-stack development

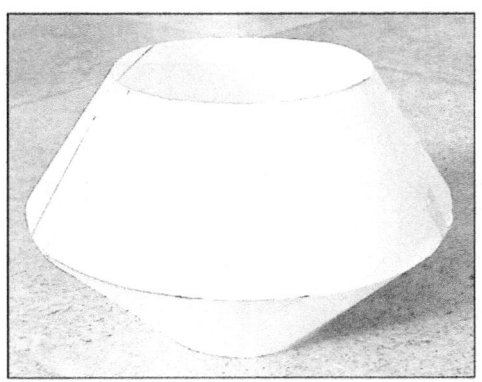

Plate 14. Diamond-stack card mock up

A trial version was made of card, see Plate 14, to check dimensions before the two metal cones were rolled to shape and spot welded. Another collar was attached that was a sliding fit for the top of the chimney tube.

The smokebox door was cut from 16 SWG steel sheet using a jig saw and finished to the line using the vertical linisher. The hinge point of the hinges need to be beyond a tangent to the door curve or the door will not open. A single closer handle was decided upon rather than the six or so clamp type retainers sometimes employed since it was anticipated that frequent easy access to the smokebox would be required to top up the fluid in the smoke unit.

Porter locomotives have a prominent bell atop of the saddletank and between the chimney and steam dome. Tubby dash is no exception and it was finally decided to keep to the monochrome décor and fittings by having a nickel-chrome one. It is 3 ½" diameter at the mouth. See Plate 15.

Plate 15. Nickel-chrome bell

The swinging mount was bent up from 12mm X 3mm black iron flat. There is no clapper as it was found that this particular example clanked rather than rang when struck. The on-board electronic bell is much more realistic.

Another striking feature of Porter locomotives was the large headlamp. See Plate 16.

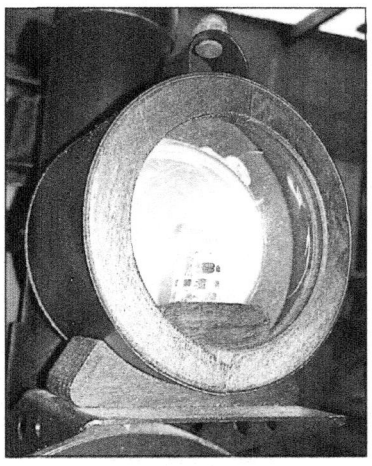

Plate 16. Headlamp

This one is 3½" diameter and fitted with a 'condenser' lens from an old 35mm slide projector by means of a turned birch ply retaining ring. An aluminium reflector, the bottom of a 'Coke' can, is attached to the birch ply rear closer. Light comes from a 12V X 1.2W LED bulb that gives a good beam for a current of only 100mA. The headlamp unit is mounted on two stepped-forward folded up bracket. Two holes were punched in each of the risers to mimic the appearance of the original.

Controls

Almost hidden from casual view at the back of the cab, the firebox houses the motor control system. The box is formed from a 6" wide X 20 SWG steel strip with a 3" radius top and straight sides making it 6" high. At the front is a ¾" plywood former to which it is attached with 3mm X 16mm CSK wood screws.

Plate 17. Firebox control panel

The rear closer is the control panel, again of 20 SWG sheet, attached by 6BA CSK screws and right

angle brackets bent up from the same material. The layout of the controls can be seen in Plate 17.

A standard 'Hall effect' electric scooter throttle was modified to be operated by a ¼" spindle and brought out to the front of the firebox panel. A regulator lever moves this over an arc of thirty degrees or so. The first trials of the locomotive quickly showed that it was uncomfortable reaching into the cab from the driving seat to hold the regulator open against the spring. So a modification was carried out to allow a remote throttle to be plugged in overriding the lever operated one. This took the form of a 3.5mm jack socket with switched contacts inserted in the wiring such that the original throttle still worked when the remote one was unplugged. See Plates 17 and 18.

Plate 18. Firebox control panel rear

The basic motor control circuit is as follows. Leads from the locomotive 24V battery are fed through a hole in the cab front just above the firebox and terminate in a flying female connector. A male connector then feeds the key switch, 24V fuse, charger socket, 24V battery indicator push switch and the electronic motor control unit. The purpose of the flying plug and socket external to the firebox is so that another 24V battery can be easily substituted for the locomotive's on board one if necessary during prolonged running. See Plate 19.

This box was simply made up of some offcuts of oak wooden floorboards screwed together and slotted ⅛" using my circular saw bench to take the bottom and lid.

Plate 19. Spare battery box

The output of the control unit is fed to the motor via the Forward/Off/Reverse switch. See Appendix 1. An unsuccessful attempt was made to find a key switch with dual switched contacts so that the 24v and 12v circuits are isolated by one key turn. A possible alternative would have been to fit a relay but this was decided against on the grounds of extra complexity. In any case all the 12v circuitry is controlled by visible toggle switches on the cab control box. See Plates 20 and 49.

Turning to the 12V system. The cables from the 12V battery go directly to the 12V fuse, 12V charger socket and the 12V LED battery indicator push switch, all on the RHS of the firebox panel.

Originally, three six-core (burglar-alarm cables) ran from the RHS rear of the firebox upwards to the cab roof then back to the cab control box. Sadly, after much careful wiring had been completed disaster struck. The final connections were made and systems tested in the workshop only minutes before Tubby Dash was to have his first public outing. Opening the throttle to set off with the first trainload of expectant passenger, there was silence. Not a whistle or a bell. No chuff. Enthusiastic parents, who were unaware of the failure, were heard to supply various 'train' effects as their little ones rolled by.

Investigation afterwards quickly determined why there had been no audio. The output stage of the 'Mylocosound' generator card was burnt out. After some thought it was decided to strip out all the effects wiring and adopt a different approach. As so often happens in our hobby, it is only after a build is completed that it can be seen that there are better ways of doing things. Hopefully this is one of the benefits of this book, the reader is able to profit from the author's mistakes. Three main issues were identified. The first was that the locomotive uses several different battery supplies for the various functions, 24V, 12V and 9V. Care has to be taken that there is no possibility of cross-connection between them despite, for example, the 24V motor supply having an input to the sound card (9V supply) to give it load information. Or again, the axle-mounted reed switch assembly needing to supply trigger pulses to the chuff input of the sound card but also fan pulses to the smoke generator running on the 12V supply. This latter hazard was simply solved by fitting two independent reed switches operated from the same magnets, thus completely isolating the two functions.

A second issue was that although the burglar alarm cable was very suitable for all the low current switching circuits it was too light for the heavier current applications such as the smoke generator described in a later chapter. Therefore to make a 'proper job' of the wiring allowing for the differing requirements, including future maintenance, it was decided to make a two-section wiring loom similar to that found in the family car. The loom having the advantages of allowing different wire sections to be combined and to drop off where required. All that was needed was burglar-alarm wires and the cores from some four-core split open 0.75A mains cable. A neat result was achieved using a 15mm x 15m roll of wiring loom tape purchased on eBay. The two sections are from the cab switch control box to the firebox and from the firebox to the various functions, described elsewhere, at the front of the locomotive. Appropriate flying plugs and sockets were fitted where required.

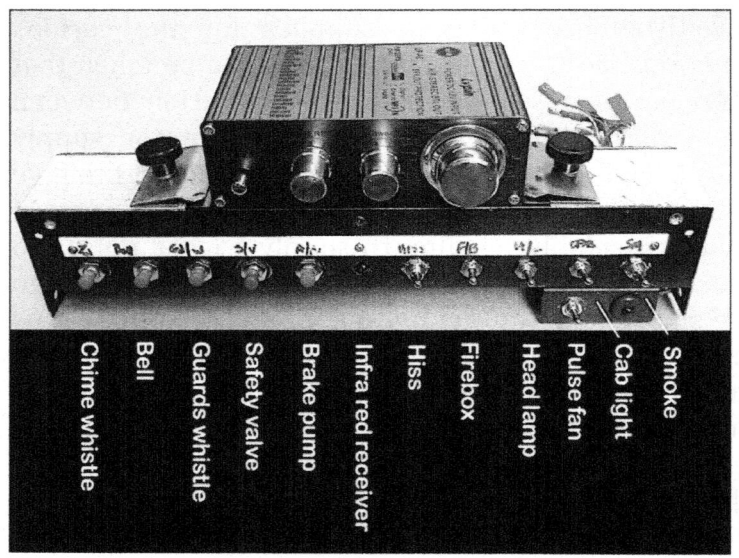

Plate 20. Cab control box

The third issue was fundamental to the whole ethos of Tubby Dash. He just had to give that indefinable 'buzz' that comes from the reciprocating motion-work and the hiss and chuff of steam. Well, despite previously believing that in a quiet garden setting the sound card output would be sufficient, it was not. A nice blue inexpensive 20W + 20W stereo amplifier, operating on 12V, was purchased on eBay and installed on top of the switch box in the cab as seen in Plate 20. Now you can hear Tubby Dash coming.

The switch box is 1¾" X 1¾" X 10¾" long and was folded up from 20SWG steel sheet to a 'U' and fitted with a ply bottom. It was made to fit under the cab roof between the cab sides. This position for the switch box was chosen to make the various effects as accessible as possible to the driver. The array of controls is hidden from the view of observers by a 'non-macho' black cloth skirt.

There are five push switches, six toggle switches and an infra red receiver on the front panel. The sixth toggle switch was an 'add-on' when the smoke unit was changed to a home-made one that is described later. A strip of white plastic above the switches provides a label space for the effects. The 'Mylocosound' card fitted in the box can be switched between British and American modes using an infra red remote control. This changes the function of some of the push switches, for example from a long whistle to an American chime whistle or another from a guard's whistle to an 'all aboard' vocal.

To the RHS of the panel are six toggle switches controlling the sound card, firebox lights, headlamp, cab light, smoke unit and pulse fan. Placing the sound card in the control box meant that it only required motor monitoring and 'chuff' trigger inputs and an output to the loudspeaker at the front of the locomotive. All the effects switching is local in the box. A rechargeable Nickel Metal Hydride 9V PP3 battery is also installed in the box. This is to maintain the background steam 'hiss' when the motor volts fall too low, as when stopped.

Cab lighting is from a 'festoon' LED mounted behind the switch box. The fire effect behind the firebox door, seen in Plate 27, was created by means of one steady red LED and a flickering yellow one taken from an electronic 'tea light'. These are mounted in an aluminium housing found in the junk box. See Plate 21. This arrangement was found to give the most realistic firebox glow.

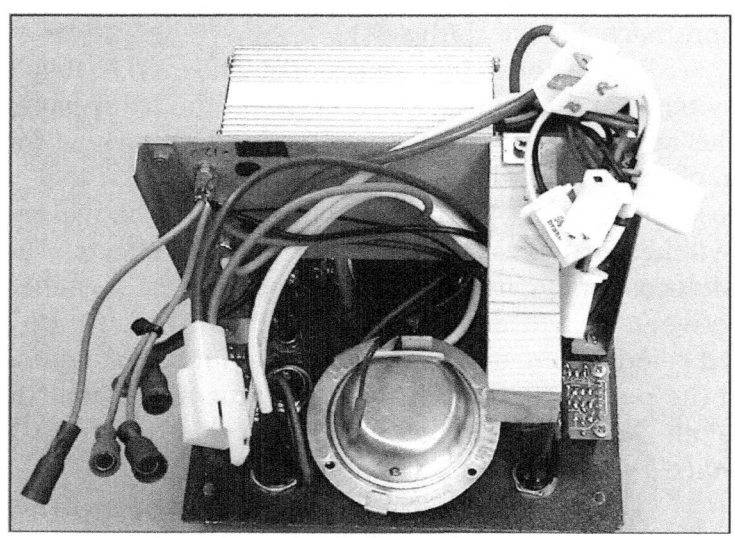

Plate 21. Firebox lighting effect housing

A pair of wires run in the loom to the smoke box for the LED headlamp mounted on top. Another pair also run via the smokebox to the 'diamond' smokestack at the top of the chimney to power the pulse fan of the smoke unit. All cables passing through the smokebox are fitted with 2 pin JST flying plugs and sockets to facilitate the removal of the saddle tank/smokebox unit to give access to the batteries.

A plywood loudspeaker box was made to fit between the frames housing an eight Ohm speaker from an old radio. Volume was found to be more than adequate with the 20W stereo amplifier. No need now for fond parents to vocalise appropriate train noises.

Smoke Unit

In order to achieve a near steam locomotive appearance it was considered essential to have 'smoke' issuing from the chimney stack. Ideally there would be puffs synchronised with the chuffs from the sound unit

Plate 22. Axle mounted magnet disc

As mentioned in an earlier chapter describing the rolling chassis, provision was made for this by fitting an axle-mounted brass disc with four magnets round the periphery. See Plate 22.

The magnets removed from the plastic case of a burglar alarm window/door switch are attached to the brass angle brackets by heat-shrink sleeving. Two of the matching reed switches mounted via their plastic cases on a bracket then give independent switch outputs for the sound trigger and for the smoke-unit pulse fan. The purchased audio chuff effect worked well using this trigger arrangement. The pulsed smoke effect turned out to be a more complex matter altogether; involving a heating element, 'smoke' fluid, and a motorised fan.

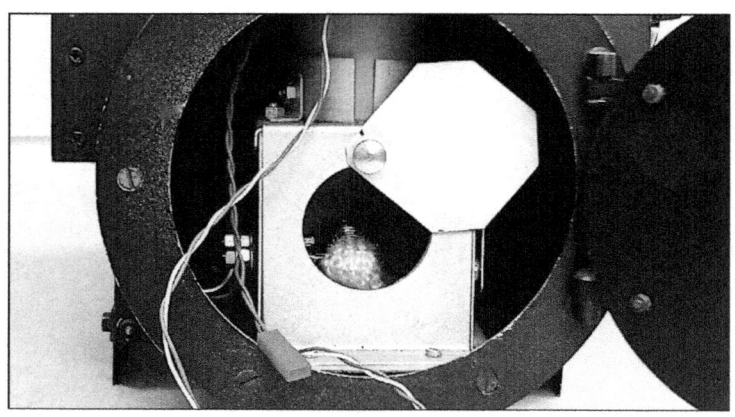

Plate 23. Smoke unit in smokebox

Many experiments were carried out with various second-hand commercial smoke units designed for model railway locomotives up to 'G' scale. None were able to produce a sufficient volume of 'smoke' for a 7 ¼" locomotive, so in the end one was constructed following the same basic principles. See Plate 23.

Some suitable resistance wire was obtained by purchasing the least expensive hair dryer sold by Argos. Once dismantled, the wire was unwound from the heater element and straightened. This was achieved by clamping one end of the wire in the vice and then pulling a length through a small hole in a piece of hardwood.

Experiments showed that reasonable smoke production resulted with about 2A current consumption from the 12V battery. Using a multimeter, a length was measured off that read around 6 Ohms. A search of the scrap box produced a 'wick'. This was a short piece of glass fibre rope left over from replacing a wood burning stove window glass. The wire was wound round the middle of the wick and mounted such that both ends of the wick dipped into the fluid tray. See Plate 24.

Copper wire stubs were crimped to the nichrome wire ends with short lengths of copper small-bore tubing. In turn these were soldered into drilled holes in 4BA brass screws. These passed through lengths of fiberglass sleeving in holes in the wall of the smoke unit box and with brass nuts formed insulated terminals for the supply cable.

The smoke unit box was folded up from the usual 22 SWG steel sheet and provided with a front door to replenish the smoke fluid. To one side is a shielded air vent and on top is a collar to fit a length of 1 ¼" plastic waste pipe to conduct the smoke up the chimney.

A toggle switch on the cab control box, See Plate 20, switches the 12V supply to the heater via

Plate 24. Smoke unit heater and wick

0.75mm wires in the loom and a 2 pin JST flying plug and socket to allow removal of the box. A red LED indicator reminds the driver that the smoke unit is on. This is especially useful when night driving since the smoke may not always be visible and 2A is being drawn from the 12V battery.

Commercial 'smoke' liquid can be purchased but a 'home-made' mixture that was found to be satisfactory resulted from one third glycerine (home baking supplies) and two thirds water. My smoke unit was installed in the smokebox where the opening door gives easy access to replenish the smoke fluid when necessary. It was intended to produce synchronised 'puffs' by incorporating a reed switch in the supply to the fan motor fitted in the 'diamond' stack at the top of the chimney. The reed switch, like the one producing the trigger for the sound unit, is operated by the four magnets mounted on a disc on the locomotive front axle. These switches and magnets are from window/door burglar alarm fittings.

Natural convection carries the smoke up the tall chimney and works well when the locomotive is stationery. Once under way however the smoke tends to be dissipated thinly.

Plate 25. Smoke unit, plastic duct and fan

In an attempt to get synchronised puffs, further experiments showed that a small fan fitted in the diamond stack and switched by the second reed switch worked well at slow speeds. See Plate 25.

There were several reasons why the fan had to be small. The first was that the fan/motor combination had to have a low momentum so that it could start and stop rapidly to produce discrete puffs. Secondly unless a relay was to be employed, introducing further complexity, the switching current of the motor (many times its run current) would greatly exceed the reed switch rating. Lastly, it was thought the motor/fan combination had to fit inside the diamond stack chimney top.

Plate 26. Tubby Dash smoking in the station

Some success was achieved with this arrangement employing a small model helicopter tail rotor and propeller. However, it was not long until condensation of the 'smoke' fluid on the fan blade and motor had gummed them up to the point where the start current burned out the reed switch contacts.

A future modification to avoid this condensation problem will be to move the fan to the position of the side vent on the smoke unit box.

In the mean time the smoke gently billowing

from the stack when stopped in the station has produced favourable comments and requests to see the fire. See Plate 26. The glow from the opened firebox door is quite realistic, particularly in the evening time, and adds a little extra to the panache of Tubby Dash. See Plate 27.

Plate 27. Glowing firebox

Driving Trolley

Whereas my tram, 'Bushmills Bluey' was a ride in locomotive Tubby Dash was not and so a driving truck was required. There were two benefits. The driver would not be taking up space on the passenger coach and an effective brake could be incorporated. Although the locomotive was all but finished, due to the pressure of time now that a railway had to be built as well, it was reluctantly decided to 'buy in' wheels and axles.

The wheels duly arrived and the parcel seemed inordinately heavy. The reason was immediately obvious; the wheels were far larger than I thought I had ordered. Investigation showed that this was due to a gross imperial/metric conversion error on my part. Well, NASA had the same problem didn't they? When I got used to the idea, 6" diameter wheels seemed fine and the driving truck construction could proceed.

Tests were carried out on the workshop floor to see the minimum wheelbase that could be used whilst retaining good stability. Eventually 16" was chosen for the wheelbase and an overall length of 30" for the truck. The same basic wooden chassis was used as on my tram, two longitudinal members 30" X 1¼" X 2" high supporting four 5/8" ID pillow-block pressed steel bearings. Four pairs of 'tight' 10mm holes were drilled to suit the 68mm bearing mounting centres and 10mm OD steel tubing pressed in place. The tubing acting as sleeves for the 8mm X 90mm sliding bolts carrying ¾" X 12 SWG springs, centring 'cups' and the bearing mounting nuts and locknuts. See Plates 28 and 29.

Plate 28. Wooden chassis members

The cups were sheared to size using the bench guillotine and the corners bent up to form the centring cups using a home-made 18" bender. This useful machine was built from a design in Newnes *Practical Mechanics*. See Plate 30.

Plate 29. Springs and cups

Plate 30. Home-made bender

Plate 31. Driving truck deck

Timber from an old discarded divan-bed sawn to a 2 ¾" X 7/8" section was used for the deck, and screwed together by 5mm X 50mm CSK wood screws. See Plate 31.

At the front of the truck a 22" length of ¾" OD steel tubing fitted with rubber feet provides footrests and the front coupling point. A 12" piece incorporates the rear coupling point. Both are bolted to the truck with four ¼" X 1¾" coach screws. See Plate 32.

The 12" X 12" X 13" high box seat has a hinged lid and was purchased as a 'flat pack' from a B & Q store. It has proved to be useful as a repository for the driver's hat and other running equipment.

Tubby Dash is not fitted with a brake and indeed one would not be very useful, except as a parking brake, due to the limited weight. So, some thought was given to fitting an effective one to the driving truck. An essential requirement when pulling passengers, as sticking out both feet in an emergency 'doesn't cut the mustard'.

Plate 32. Driving truck

Over the years many different braking systems had been installed in my trams and locomotive driving trucks. Some were better than others and all seemed to need quite complicated linkages to be effective. To be honest I had more interest in going than stopping and so brakes were a necessary but rather tiresome extra. The one that evolved is simply a hinged piece of ¾" plywood forced down on the rear truck wheels by a cam on a cross shaft rotated by an outside lever. This arrangement has proved to be very effective. See Plate 32. An upper 'spud' providing a parking brake position.

Plate 33. Parking brake

All was well until a derailment problem appeared when the driving truck was negotiating a chicane in the track. After some investigation, it was concluded that the driving truck springs were too short and 'strong' even for my weight. There was insufficient compliance to cope with certain track irregularities. A new set of longer 'weaker' springs was installed and now all appears to be well. Although the lever operated brake still functioned perfectly with the driver on board, now it was useless as a parking brake due to insufficient throw of the operating cam for the longer springs. So, the opportunity was taken to install a new and more railway-like parking brake. See Plate33. This consists of an oak shoe pressed downwards onto one wheel via a ½" Whitworth threaded rod. It turns in a threaded brass nut by means of a cranked handle.

Yes, it does rain a lot here in County Down due to apparently endless southwesterly fronts coming in off the Atlantic Ocean. Nevertheless, it seems to be a very rare occasion indeed that one can organise any outdoor event that is not spoiled by rain piddling, Jimmy Riddling (or whatever you like to call it) down as soon as steam is raised, or in the case of Tubby Dash the hiss is switched on.

Having suffered the ignominy of 'getting a push' on the steeper gradients of an aluminium track after a shower of rain had seriously diminished traction in the past, some thought was given to averting this distressing situation.

The thinking went like this: If the wheel slippage is caused by damp, why not dry the track? The simplest and most inconspicuous way to do this seemed to be by rubbing the track rails with pads to clear most of the surface water. See Plate 34.

Two slices of household 'Sponge scourers' were mounted between large woodscrews on a hanging plywood flap under the driving truck. The device is retained up out of contact with the track, when not in use, by a suspension wire. A marked increase in performance was noted (imagined?) after a shower of rain and a circuit with the 'dryer' in use. Could it be that the top surface of the track was now reasonably dry for adhesion but the flanges were being lubricated allowing negotiation of the 15' curves with greater ease? Bring on the rain?

Plate 34. Track dryer

Windy Hill Railway II

There are probably few places on the planet that have had four railways built in the same garden. The story of the building of the first three of these was told in *One Man's Garden Railways*.

In 2009 we had sold our home of thirty-five years, including its railway, and downsized to a nearby bungalow with a small garden. Eventually a compact 7 ¼" gauge railway was built there.

My previous book, *One Man's Garden Railways*, concluded in the following way:

Sometimes, alone, on a summers evening, 'Bluey' would be brought out of the shed and a quiet pint or two of home-brewed beer would be enjoyed as we rumbled around between the fronds of garden foliage. Perhaps another set of points inserted just here, might switch us to that other time-track where we would always 'live the dream'?

Well, the points never were inserted but despite this, the switch was made to another track in a most unexpected manner. It happened like this:

One day when out for morning coffee my wife Ann and youngest son Alan were passing a local estate agent's window. They were amazed to see that our old home was back on the market after some six years. It was incredible. The young couple that had bought it had such plans for the future. It was a turning point in life such that the following weekend was spent in wonder and disbelief. Alan, who had not been in a position to buy earlier, was in a daze. How could he work this? He now had a town house that was nearly completely renovated but still needed many finishing touches to make it saleable.

Plate 35. Mini digger at work

We will pass quickly over the stress and anxiety of the next six months. Suffice to say that the saga of dealing with house agents, solicitors, building control inspectors and jobbing builders is something that we would rather forget, but never will. He did it; he bought it in July 2015. It was 'anastasis', a rebirth. Soon eldest son Andrew was talking of a late summer garden party. It was a great success. Christmas followed with a memorable family get-together in our former home. In the New Year, 2016, I started to build Tubby Dash and in the spring there were discussions about the creation of another 7¼" gauge railway in Alan's garden. This would be my sixth railway and the fourth at this venue. By the tenth of May a mini-digger was at work in the old orchard. See Plate 35.

Plate 36. Tree roots

The route of the new line is a modification of the old 5" gauge 'Windy Hill Railway' and of a similar length. Excavation of the track bed was to a depth of around 9" or so depending on humps and hollows in the ground level. Because the area had at one time been an orchard there were many tree roots pulled up by the digger bucket creating patches of loose soil. See Plate 36.

Plate 37. Ballast filling

The shallow trench was lined with black heavy-gauge builder's polythene that had been cut into one metre wide strips and liberally punctured with a garden fork to allow drainage. Many barrow loads of 15mm stone chippings were wheeled by my two strapping sons from a pile in the house-yard to the track and roughly levelled using a rake. Note the garden seat for weary ballast layers in Plate 37. Eventually, a sharp knife was attached to a wooden lath with insulating tape and drawn along the route slicing off the excess polythene at ground level.

All the existing Rocklands Railway track at the bungalow was lifted and reused so that only one pack of 2.5m X 28 lengths of new rails had to be purchased. The new track work was laid using the same 16mm high extruded aluminium rail section. This was screwed to 14" X 1¾" X 7/8" keruing sleepers by 5mm X 25mm pan-head stainless steel screws and washers. On completion, more chippings were added and raked in as holding ballast. The old points were also incorporated to give a siding through a new entrance into the workshop, now christened the Railway Room, for storage of the trains.

Plate 38. Old sleeper

An interesting 'archaeological' find turned up during the track excavation. Some of the old 5" gauge slotted sleepers appeared, still in perfect condition, after thirty years or so buried in the ground. Thus illustrating the efficacy of creosote immersion. See Plate 38.

Summer Days on the New Railway

A railway needs a station as a focus for intending passengers and as an arrival and departure point for the trains. In a garden setting it also provides an opportunity for a decorative structure supporting flowering plants and lighting.

Plate 39. Rocklands Station

Fortunately, the one built for my previous Rocklands Railway still existed. Since the latest family car replacement had no tow hitch as yet, the station was sawn along a horizontal line using a jigsaw so that the two bits would fit into the hatch-back. Four 'slater's lath' splints restored the structure to its previous appearance but now as a prominent feature of the Windy Hill Railway II. See Plate 39.

Plate 40. Routing station nameplate

One casualty was the station name board, which had been dumped since the Rocklands Railway had been consigned to history. Luckily the home-built pantograph rig was still extant along with the set of master letters that had been used to rout the original name board. The masters were produced by printing individual letters on paper from a suitable typeface, in this case Gill Sans, using an inkjet printer. A scroll saw was used to cut out the letters after they had been stuck down to ⅜" plywood sheet.

At first it had been intended to rename the station because of its new location. However, it was felt that there were several good reasons to keep the existing one. Windy Hill Railway II is located in Carricknadarriff townland in County Down, Carricknadarriff being the anglicised version of the Irish name meaning 'Rock of the Bull'. So a rocky landscape seemed appropriate and kept a sense of continuity with the past. Using the original master letters also had the benefit of saving tedious hours of work to make different ones. See Plate 40. The blank being routed is an offcut of oak T & G flooring.

The new Windy Hill Railway (WHR) was only completed two weeks before the garden party. Some irregularities in the track levels that were causing coach derailments were corrected, and at long last Tubby Dash was tested pulling several adult passengers riding on the 'Ecocoach'. It was a considerable relief to find that traction was adequate to negotiate the rather steep gradient to the summit of the line. My old tram, 'Bushmills Bluey', had no trouble of course, having the adhesive weight of me as driver. However, Bushmills Bluey does not have the panache of the new dark stranger, Tubby Dash.

We were blessed with a warm dry evening on the thirteenth of August. When the daylight faded the benefits of our preparatory work began to appear. As night fell the tracery of lights and the blazing fire pits created an enchanted scene with Tubby Dash, headlight on, weaving his way amongst the groups of people with a train of happy children.

Plate 41. Summer barbeque

We operated more or less to a time-table and a young girl called Petra sat at a the station entrance issuing tickets. It was only after some time we realised that she was charging for them at the rate of 1p a go. It is rumoured that she made 56p even though she let 'family' on free. It was all part of the 'colour' of the evening.

Although it had required a great deal of work in a short space of time, there was now a permanent railway as a basis for future occasions. From now on Tubby Dash was to feature on various occasions in hauling short trains of young passengers past various attractions. Halloween is well catered for when a sort of modern ghost train is operated using simple technology for some of the effects.

Referring to the track plan in Plate 42, each journey commences at 'Rocklands' station, which is backed by a high overgrown earth bank. During spring and summer days there are colourful displays of wild flowers that include Comfrey, Stitchwort, Cow Parsley, Daisies, Buttercup, Dandelion and Cuckoo Flower on this south-facing slope. See Plate 42. The station is situated on a slight downgrade making an easy start for loaded trains, especially when the aluminium rails are slippery after a shower of rain.

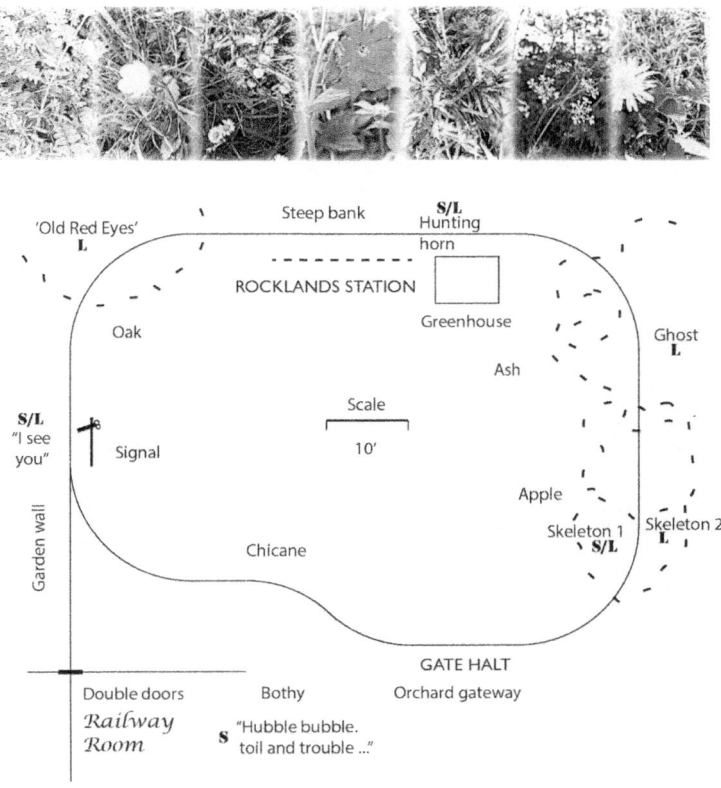

Plate 42. Windy Hill Railway II track plan

Plate 43. Granddaughter Sophia, deputy driver

Plate 44. Miniature Co. Donegal Railways signal

Plate 45. Tubby Dash in the sun

Plate 46. Deputy driver and self

Plate 47. Tubby Dash and Bushmills Bluey

Plate 48. Leaving Rocklands Station

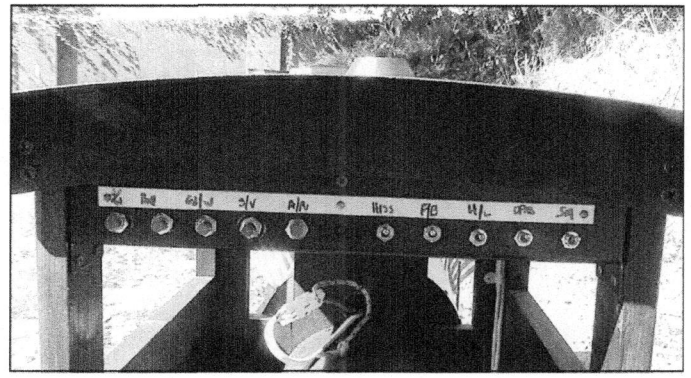

Plate 49. Cab control switches

Plate 50. Summer barbeque sign

Hedge-Conditioned Home Brew

There has been occasional mention of hedge-conditioned home brew beer in previous chapters. It might even be mentioned again, so perhaps a word or two explaining this bit of romantic nonsense would be in order.

The phrase originated from having insufficient cool storage space in the family home back in the 1960s. This was long before the first Windy Hill railway was built. The crown-corked pint bottles were stood in the long grass under a convenient hedge. On many a long summer's evening in these northern latitudes the snake-like hiss of the opening of another bottle of this 'hedge conditioned' beer could be heard. Under the circumstances, it is perhaps understandable that the said bottles acquired a legendary reputation. Jumping forward fifty-five years, my eldest son Andrew was extolling the nectar-like qualities of 'hedge-conditioned' home-brew in some expensive hostelry in Belfast's exotic Cathedral Quarter. A disbelieving fellow-drinker admitted there might be something in it when Andrew assured him it "was under a *Hawthorn* hedge."

With the repossession of the old homestead the practice has been resumed as seen in Plate 51.

Plate 51. Hedge-conditioned home brew in 2017

A canopy of deep green hawthorn leaves allows a little dappled sunlight to gleam on the dark glass of the beer bottles. The damp grass and undergrowth embrace the pints like a clutch of eggs in a nest. A cooling breeze wafts through the hedge. Taken together these natural attributes provide a hundred and fifty year old home-brew conditioning facility. By extension, and after sampling a couple of bottles, the thought arises that there are thousands of miles of hawthorn hedges in the British Isles, second only to oak as a native species. If even some of them were in use as stores for home-brew beer there would be, as we say in Ireland, a 'quare' supply.

The Railway Room

The Railway Room was adapted from a disused workshop. It has 4½" thick concrete block walls and a sloping corrugated-iron roof.

Plate 52. Railway Room entrance

Seen here on a damp autumn day is the old entrance door with a few of the railway signs created on my computer, ink-jet printed and laminated in plastic sleeves. After some months the ink is already fading and a future project is to make new ones using an inexpensive CNC laser engraver from China. See Appendix V.

Plate 53. Double entrance doors

For a railway-like appearance it was decided that the other new entrance for the railway track would have to have doors of the double centre-opening type with a curved top. Each half of the pair was ledged and braced and was constructed from 18mm pine tongue and groove boarding. The curved top or 'head' was formed from several layers of boarding, jig-sawn to shape, and thin plywood bent round and fixed with wood screws to give a solid looking arch. Internally the doors are 'locked' by a wooden lath dropped into place between steel brackets on the doors. A simple method rooted in antiquity seen on barn doors in the Mourne

Mountains in Co. Down and no doubt in many other places. See Plate 53.

Initially a short section of removable track was fitted to allow the doors to be opened and closed. However, it was found that due to increasing use of the railway, sometimes for a quick trip on a dark and windy night, something more convenient was required. In the past, on a previous railway, a flap was employed. Experience had shown that this method, although effective, suffered from a major problem that occurred when visitors opened the door without first raising the flap. The method shown in Plate 54 has been adopted as an alternative and hopefully will be mouse proof.

Plate 54. Track door detail

It was quite a puzzle as to how to lay out the storage tracks in the Railway Room to maximise the use of space. At some point in the distant past I had been greatly taken by David Lean's classic black and white film *Brief Encounter*. It was a wonderful tour de force with the billowing clouds of steam, the war-time subdued lighting and the impossible romance in the station tea-room, all set to Rachmaninoff's 2nd piano concerto. I might have had a couple of pints of 'hedge-conditioned' home-brewed beer at the time, but I just had to have a 1940s style tea-room with a table and chairs and low-level lighting. Fortunately, the romantic aspect was fully taken care of by the love of my life, Ann.

Plate 55. Railway Room bar

The Railway Room has a small bar equipped with four LED down lighters that add a bit of life to the glassware and pints of home brew. They are controlled by a switch under the bar roof. The 12V supply is from wiring that was installed in the late

1980s. The Railway Room was previously an engineering workshop where I made architectural metalwork. Follow this link to see it when weathervanes were being made there.

 https://youtu.be/7WFr91P4Sgs

There were so many interruptions to the mains electricity supply in those days, due to storms disrupting the overhead high tension supply lines, that a 12V Lucas 'Freelight' wind generator was installed to feed the workshop and the dwelling house. Nowadays, there is just a 12V car battery and a mains charger. The curved and ribbed roof is from a now disused bogie-coach. It has recently acquired a 'thatch' of dried hops complete with two resident birds, one called Fred nesting in the nearest corner. See Plates 55 and 56.

Plate 56. Bar bird Fred

Left of the 'Aviator' poster, a door leads to a very small kitchen with a mains water tap and sink unit. It is a hive of tea-making activity when seasonal railway parties are in progress. There is a small multi-fuel stove that can be lit if the weather turns really cold. Over to the right of the bar is a Phillips 'Ambilight' TV currently displaying a Halloween theme. For Christmas time there are various forest snow scenes complete with a compilation of background steam locomotive sounds. All were acquired from YouTube and edited in iMovie.

Tubby Dash and driving truck are parked directly behind the bar and behind them at the wall is 'Bushmills Bluey'. The shelf above holds the 24V and 12V battery chargers for both locomotives.

Plate 57. Phoenix

To the right of the bar are the railway entrance doors and in front of them, on the shelf, is my 3½" gauge Phoenix. See Plate 57. This locomotive was built from measurements taken from the original in The

Folk and Transport Museum, Co. Down. Phoenix ran for a quarter of a million miles on the three-foot gauge tracks of the Co. Donegal Railway. It is 24V battery-electric powered, using many of the same electric scooter components as Tubby Dash and Bushmills Bluey. Before Phoenix was finished, it can be seen giving an 'electrifying ride' on the MESNI track at Cultra.

 https://youtu.be/Vh1dCcx9XK0

Turning further right is the main source of winter comfort. The stove with its tall black-enamelled chimney provides surprising warmth in

Plate 58. Railway Room stove

such a large non-insulated area. See Plate 58.

There was insufficient space to incorporate reasonable radius points to allow access to two storage roads in the Railway Room. So, one track was laid through the door such that a loaded train could be driven in as when escaping from a shower of rain or just at the end of the day. Another parallel track between it and the side wall gave storage for currently unused vehicles. Timber boarding of ¾" thickness was screwed to the sleepers between the rails and the tracks making it possible to manually shift the bogied trucks and "Bushmills Bluey' tram across to the running track without the wheels becoming jammed in the declivities. For this operation it was found that provision of 'bogie locks' (4" nails inserted in holes in the floor of the trucks and mating with others in the bogie decks) greatly assisted in this operation.

Out on the track, it can be quite difficult to align the front and rear bogies of a ground-level passenger truck especially on a curve in the dark. They are often hidden under the body, sometimes requiring one to lie on the ground in a rather undignified posture.

An unexpected bonus is to have the use of the extra seating provided by the locomotives and coaches stored in the Railway Room. Even cold winter evenings can be enjoyed with the multi-fuel burner stove well stoked with whatever fuel is to hand.

Ghouls and Wraiths

A garden railway is well suited to running a ghost train on Halloween night, provided it is not raining of course. The natural setting has the advantage of unpredictable sensations caused by the evening breeze brushing foliage across one's face for example. A startled bird flapping noisily from its perch in a tree can occasion another frisson of fright.

Plate 59. Full moon over Co. Down

Such natural events have a disproportionate effect on a group of passengers somewhat under the influence of imaginary Halloween spirits in the darkness.

Just as in an indoor funfair, the track guides one through a series of happenings experienced at a low level in stygian gloom. To add to the potential frights that nature provides, it is a fairly simple matter to plant a few less subtle ones. Accordingly, so far, six proximity-detector operated devices have been assembled. See details of the construction in Appendix II and III. All are triggered by approaching trains or by persons walking past. Children, like cats, seem to like to 'walk the track' even in the dark. Three of these units bring on a 5W LED lamp along with an audio recording between two and twenty seconds long. The other three are lighting only.

With the stage set we will brave the ghouls and wraiths and set off on a journey on the Windy Hill ghost train.

On Halloween night, the railway was mostly in darkness apart from occasional moonlight and some localised pools of light. With an 'all aboard' from Tubby Dash's conductor (American electronic vocal) and a warning blast from the chime whistle our first train of the evening sets off. There is a satisfying chuff-chuff and accompanying whiffs of smoke from the smokestack as our locomotive leaves the down gradient away from the station and starts to climb towards the left hand curve past the oak tree. Suddenly a large pair of red eyes stares balefully at us out of the undergrowth under the tree. They blink out as we draw away with the blackness of the garden wall on our right. Rounding the curve, on our left is a table within a tented gazebo holding a collection of exotically carved pumpkins supplied by family and visitors. See Plate 60. Last Halloween, a prize had been awarded for the 'best' effort. But this time the standard of all the entries was so high that everyone got a prize.

Plate 60. Pumpkin display

They almost look alive as the eyes and jagged mouths appear to move as the candles gutter in the evening breeze.

Proceeding towards the signal we are suddenly bathed in a bright light and a creepy drawn-out voice calls out, 'I ... see ... you'. The light somewhat diminishes our night vision and makes it seem even darker than it really is.

Our many wheels click and clack over the points that give access to the siding leading to the Railway Room.

Straight ahead is a blank wall. Last year it was used as a screen for an inexpensive video projector. This displayed a very early black and white Walt Disney cartoon of barnyard creatures and skeletons performing amusing Halloween gyrations with appropriate audio accompaniment. The film was downloaded from YouTube and recorded on a USB memory stick. The video projector was purchased from eBay for £70. It has a long-life LED lamp but has far less Lumens than other much more expensive projectors. However, the brightness is quite adequate on a dark night.

The files on the 'stick' repeat and after the cartoon there was a Halloween scene of witches flying on broomsticks, ghostly figures and monstrous bats all in front of a roaring fire accompanied by a whole CD's worth of commercial 'scary sounds'.

Plate 61. Toil and trouble

This year the projector has been moved to a position under the roof of the Bothy, on our right, as we round the bottom curve and negotiate the chicane in the track. In addition to the videos of the previous year there is now an illustrated sing-along piece "Hubble bubble, toil and trouble …". Wisps of steam swirl from a cauldron of witches brew. See Plate 61 and Appendix IV. After passing the gateway to the orchard, our large headlamp throws its beam to the left and with hearty chuffs we begin the climb under the trees.

Plate 62. Hanging Skeleton

A long male scream rends the air. Before us are two skeletons. One, backlit from the ground, hangs from the branch of an apple tree. He has red smouldering eyes. See Plate 62. The other seems to be emerging from an earthen bank to our right, lit from behind a small tombstone. See Plate 63.

Plate 63. Emerging skeleton

Climbing towards the summit of the line, a ghostly wraith materialises half concealed by tangled undergrowth. Our driver is not frightened and dispatches the Edwardian lady phantom with a short blast from the chime whistle. In daylight, the ghost can be seen to be a brush shaft with a balloon tied to the top and all gracefully covered by a trailing cotton sheet. The neck delineated by a length of cord. See Plate 64.

Plate 64. Edwardian lady ghost

Leaving the summit on the downgrade we are coasting quietly towards the station with Tubby Dash in overrun. We are lit up and given a blast from a hunting horn, perhaps nothing to do with Halloween, but atmospheric in the darkness. We slow to a stop using the driving-truck brake past dimly-seen waiting figures. We are back in Rocklands Station having survived the ghouls and witches of Halloween.

A Blink in Time

One of the many mystical qualities of home-brewed beer is that of apparently opening cracks in the spectrum of time and allowing a little time-travel. Such an occasion occurred on the wet Wednesday afternoon of 19th July 2017, when for a few hours three old friends once again boarded a train on the Windy Hill Railway and forty years fell away like an Irish mist and it was 1977 again.

Peter, Frank and I relived the glory days in the early years of BBC TV in Northern Ireland over a few pints of home-brewed beer, hedge conditioned of course, in the Railway Room and then boarded a train and steamed into the past on the Windy Hill Railway II. True, it was now Tubby Dash 'steam' rather than the 'real stuff' produced by Friar Tuck (he of the dirty brown habit) but that was no hindrance to a reprise of the original 1977 trip seen in Plates 65 and 66.

The original photo shows Peter Lindsay, at that time a BBC TV film sound recordist, at the throttle. Peter now works in the movie industry as a Production Sound Mixer, including on such Hollywood blockbusters as *Avengers: Age of Ultron*. The passengers riding 'beer-crate' trucks are Frank Murphy, another BBC colleague – now a retired Picture Editor - and myself enjoying a pint at the rear. The grease-top hats have remained the same but the beer-crate trucks were long ago returned to their original purpose.

Plate 65. 1977 trip with Friar Tuck

Plate 66. 2017 trip with Tubby Dash

Back in the Railway Room the beer ran out; and my son, Alan had to be dispatched for further supplies from the bungalow garage two miles away.

He had the wisdom to bring more than the requested six bottles. He brought a crateful, which was just about right.

Long before the existence of the Windy Hill Railway back in the 1960s when the Burney's Bog Railway was extant at our bungalow home in Glengormely, County Antrim these same friends used to visit for the purpose of quaffing another style of home-brewed beer.

In those days when the ingredients for home brewing beer used to be sold in chemists shops (health drug?) as well as other outlets, my ever-loving wife used to make the beer for me. In the impecunious days of our early-married life it was much cheaper than the commercial stuff, and was considered to be better for you.

Many a Sunday evening was spent bashing the cereal grain in a bag with a brick on the back step to crush it before boiling it in the first stage of creating the wort. After all the other ingredients were added and the quantity made up with water to five gallons in the fermenting vessel, it took about seven days for the completion of the process and no more carbon dioxide bubbles to appear at the surface.

The fashion in those days was to bottle the fermented beer in half-pints so that there were eighty bottles to cap for a five-gallon brew. This was achieved by setting the bottle on the step and placing the crown cork on the mouth with a shaped steel punch on top and smartly tapping it with a heavy hammer. The odd bottle would burst asunder in complaint at this cavalier treatment. After all this effort, patience was required for a month or so to allow the brew to reach maturity and the first ceremonial hiss to be heard as a bottle was uncapped. Mind you it was not unknown, when desperate, to dip the fermentation vat for the odd yeasty pint.

Believe it or not, there is a railway connection to this tale of home brew making. During our 2017 reunion, casual mention was made of the 'beer train' that operated on the Burney's Bog Railway at Glengormely all those years ago. Three of the assembled company corroborated each other's stories about the arrival in our lounge of an under floor goods trains loaded with clinking beer bottles.

Now, I have to state that no such train existed other than in my imagination, but quite possibly I may have mentioned the desirability of such a scheme. True there was an 'undercroft', that is a fancy name for a space under the floor of the bungalow, and it is also true that the floor was raised above the level of the five inch gauge railway in the back garden. So a branch line would have been possible, running from the storage crates in the garage to an under floor station accessed by sliding doors in the lounge.

As far as I can remember, there was indeed some beer stored under a trapdoor in the floor. But railway supplies, with my wife loading the train and dispatching it to automatically stop and ding a bell to announce its arrival; don't think so. However, three grown men (mature?) assured me that that it had been so and described the sliding doors on the cupboard giving access to the track beneath. Yes there was a cupboard with sliding doors…

When the last of the visitors cars had gone I set off on a valedictory circuit of the Windy Hill Railway

II. As Tubby Dash and I passed under the trees and through the station, now indistinct in the failing light of a damp July evening, I began to imagine how it *might* have been in that other world of long ago, around fifty years or so. The points were changed and we chuffed up the steep gradient into the Railway Room. The double doors were closed and the locking bar dropped in place. Maybe it was as they said, certainly appeals to me, home brew is amazing stuff...

THE END

Abbreviations

A	Ampere
BA	British Association
CSK	Countersunk
dc	Direct current
ac	Alternating current
mA	milliamp
mW	milliwatt
OMGR	*One Man's Garden Railways*
PCB	Printed circuit board
PIR	Passive infrared sensor
RHS	Right hand side
SWG	Standard Wire Gauge
V	Volt
W	Watt

CHARLES CARSON

Appendix I

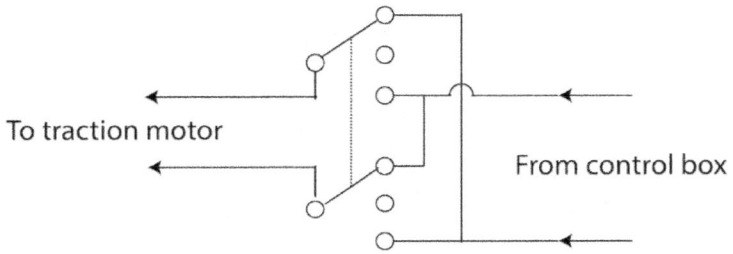

Plate 67. Reversing switch wiring

CHARLES CARSON

Appendix II

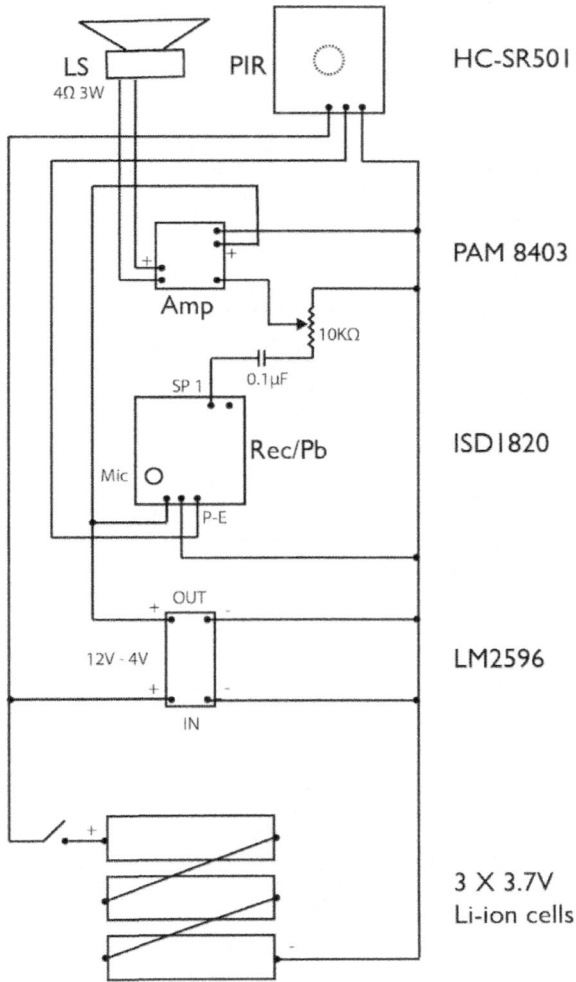

Proximity detector audio fx unit

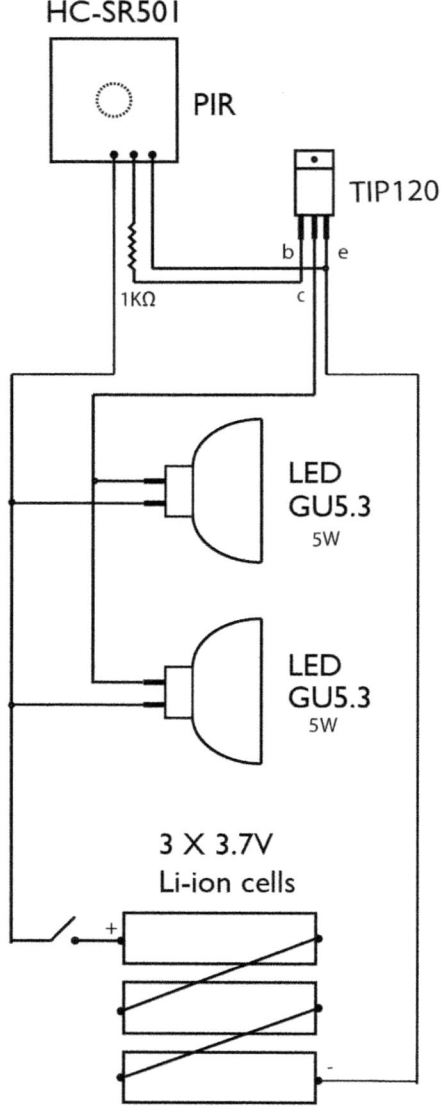

Proximity detector lighting fx unit

THE MAKING OF TUBBY DASH

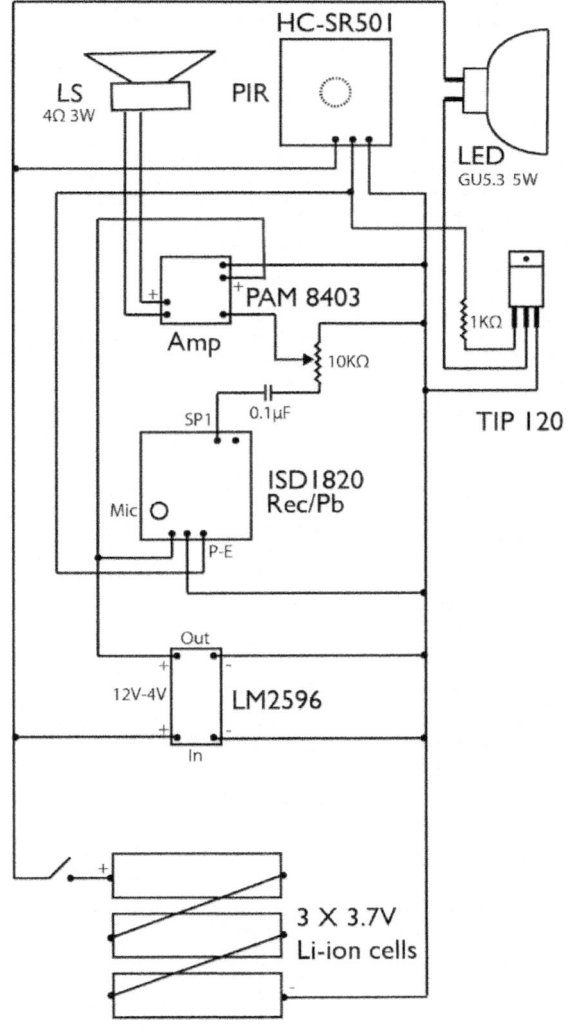

Proximity detector audio & lighting fx unit

CHARLES CARSON

Appendix III

Round apertures for the board mounted loudspeakers and LED lamps were produced by using a tank-cutter from both sides of the ply to ensure a clean hole.

Since the lighting and audio effects units will only be used for brief periods during special occasions such as birthdays, Halloween and Christmas it was decided to keep the construction as simple and inexpensive as possible. See Plates 68 and 69.

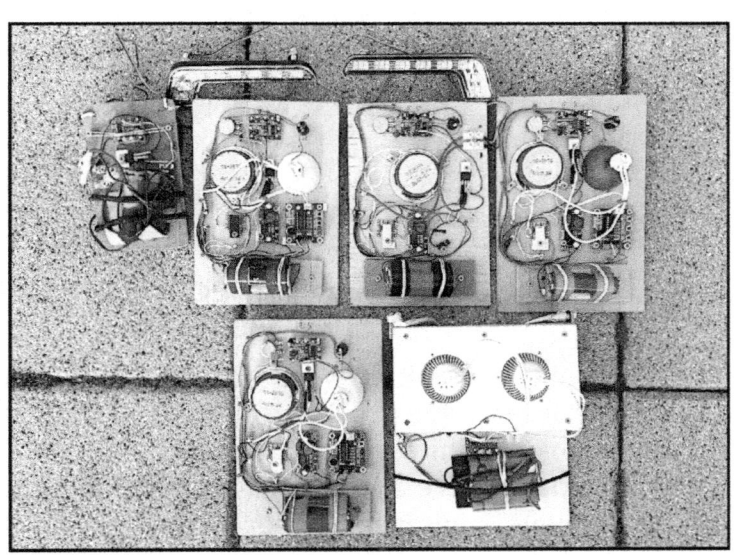

Plate 68. Six effects units

Apart from the electronic items ordered in from eBay the general construction was on 3mm ply with 3 X 12mm woodscrews with heads tinned after being screwed in place wherever required for connection points in the circuit. This combination making a sort of 'carpenter's' printed circuit.

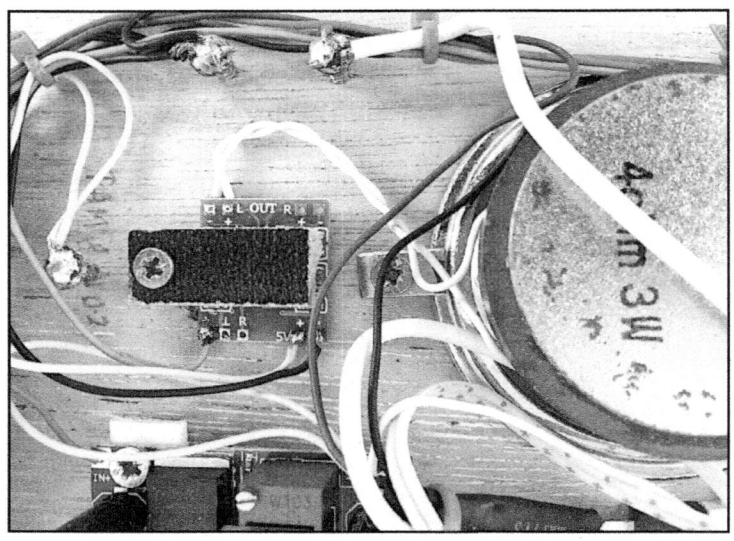

Plate 69. 'Carpenter's' PCB

The coloured wires were obtained from six core burglar alarm cable providing a good selection of colours; red, black, white, yellow, green and blue.

Essentially there are three basic units to cope with the various envisaged effects. See Appendix II and Plate 68. All are intended to be triggered by a passing train or persons walking past. For example two Halloween lighting effects were 'Old Red Eyes' using a pair of lamps behind red plastic filters and a 'ghost' with a single lamp illuminating the wraith from a distance. Another variation is a revealed skeleton with a scream using the combined audio and lighting unit. Then there could be several audio only 'scary noise' effects.

Another design consideration was to battery power the units to avoid having to run cables to each one. This proved to be practical as the current consumption was so modest and for short intermittent periods only. The batteries used were made up of Li-ion cells that have a very low self-discharge rate, indeed the batteries did not require recharging for several seasons.

Appearance of the units is hardly improved by shrouding them in plastic bags to protect them from rain, however they are fairly inconspicuous in the undergrowth at night.

Plate 70. 'Old Red Eyes' front and back

Referring to 'Old Red Eyes' seen in Plate 70. The battery consists of three Li-ion cells taped together and wired in series using car 'bullet' connectors. These make it possible to charge the individual cells to maintain their balance. The cells were acquired from discarded laptop batteries that often have only one faulty cell making them of no further use to power a computer.

Typically these cells produce a terminal voltage of 3.7V when fully charged, thus 11.1V to power the units.

Plate 71. Proximity detector board

Proximity detection is taken care of by the Infrared PIR device HC-SR501. Cost around £1.70. See Plate 71 and Appendix II.

A potential complexity arises where some of the devices on board require 12V, some 5V and others 3.3V. To avoid having different batteries to obtain these voltages (and the extra individual charging set ups necessary) it was decided to use a voltage step down board and have only one 12V battery. A pursuance of the data showed that two supplies would suffice and be within the tolerances of the devices. That is (1) 11.1V and (2) adjust the LM2596 to give a regulated 4V. Cost around £1 each. See Plate 72 and Appendix II.

Plate 72. Voltage converter board

The audio effect uses a ISD1820 device sold for use in audio greetings cards at around £2.70 on eBay. This has a built in microphone for recording and a 0.5W 4Ω speaker output. There are onboard record and playback buttons as well as trigger connections. By pressing the record button between two and twenty seconds of audio can be recorded. Scary sounds played off a commercial CD with the speaker of the player close to the microphone gives good results. See Plate 72 and Appendix II.

Plate 73. Record/playback board

Another good source of audio effects has been found to be 'ring tones' intended for mobile phone use. These are ideal as they are typically short, around three seconds or so. The speaker of the mobile phone can be held in close proximity to the microphone of the ISD1820 Record/Playback board. See Plate 73 and Appendix II.

Plate 74. Stereo amplifier board

The output from 0.5W speaker is unlikely to scare a nervous cat so it is fed to another board, a PAM 8430 stereo amplifier via a 10KΩ volume control. See Plate 74 and Appendix II. This has a 3W output. The eBay speakers used have Mylar cones making them suitable for outdoor use. Cost around £2 on eBay. See Plate 75 and Appendix II.

Plate 75. Speaker

Where the boards use pins rather than soldered connections it was found to be convenient to purchase Dupont crimp connectors and housings. See Plate 76. They cost about £2 for a packet on eBay.

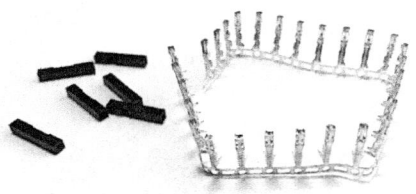

Plate 76. PCB connectors

The completed unit is protected from rain by a slipping a plastic freezer bag over it from the top and attaching a hanging loop of cord by two bulldog clips leaving the bottom open. In practice it has been found that the covering has minimal effect on the working of the proximity detector or the output of the speaker and lamp.

The PIR or infrared detector works over a good range, in fact the on board controls need to be adjusted to limit it so that only more localized personnel movement triggers an event. Without this reduction, switching can occur from movement up to seven meters away.

So far six units have been made. 'Old Red Eyes', the Edwardian ghost and the emerging skeleton are lighting only. The hanging skeleton, the hunting horn and the 'I see you' effects are both audio and lighting.
They are planted around the track in suitable positions with regard to the trees and undergrowth as seen in Plate 42.

It is intended that the proximity detector units will be used for seasonal journeys other than just Halloween. For example the audio record/replay units could be reprogrammed to perform short renditions of Christmas carols or sleigh bells and so on as a Santa Special passes miniature churches or villages in the undergrowth.

Appendix IV

A search was made for an inexpensive method of producing convincing looking emanations from a witches' cauldron. Consideration was given to various methods of achieving this. Such as compressed air blown through the liquid, a smoke unit similar to the one in the smokebox of Tubby Dash and to dry ice. All were rejected as too cumbersome or too expensive.

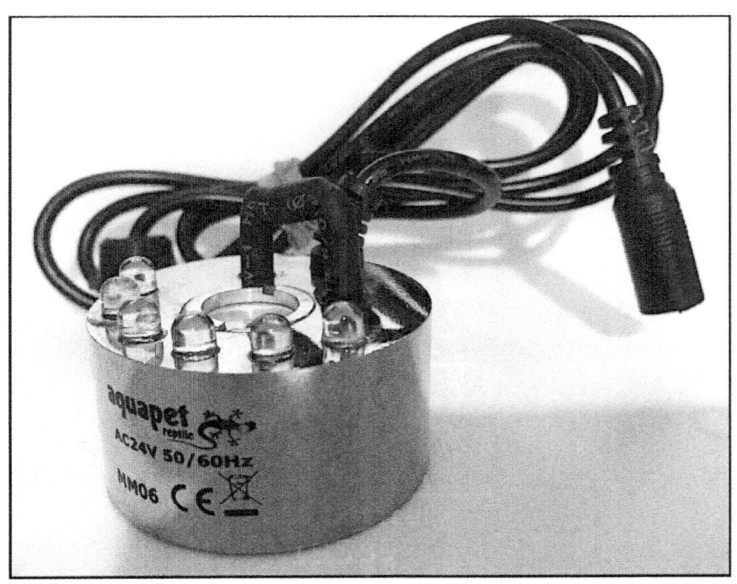

Plate 77. Mister unit

A reasonable compromise between cost and effectiveness was found in a 'mister'. No not the male kind, but a small device that uses ultrasonic sound to agitate a layer of water on a ceramic plate and produce water vapour or mist. The unit cost about £17 on eBay. See Plates 77 and 61.

This unit has six LEDs that slowly cycle through a range of colours. In the dark the illuminated mist is quite effective although perhaps not as voluminous as one would like.

Much less expensive ones are available from China if the extended shipping time is not a problem.

Appendix V

To avoid the necessity of making more master letters for my home-made router engraver an investment was made in this 2,500 mW Elekscam laser engraver and software. See Plate 78.

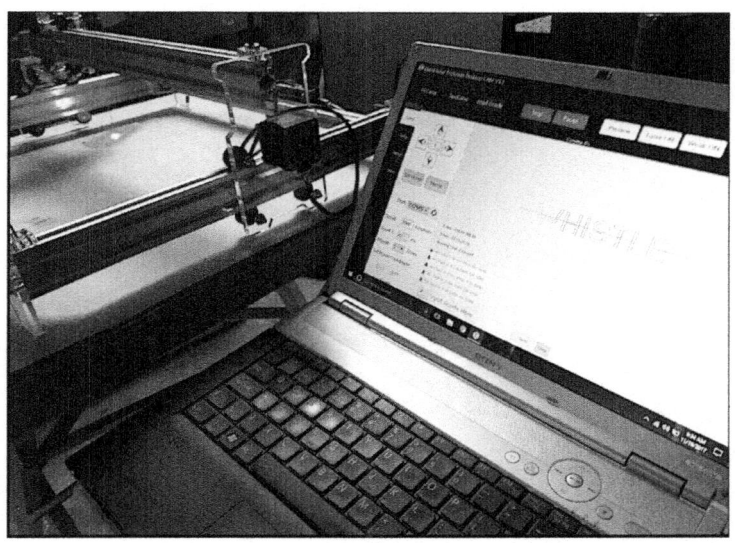

Plate 78. Laser stencil cutting

It is capable of engraving or cutting up to A3 size. Although it can only cut thin card or plastic sheet it is possible to produce good quality stencils. These can then be used to paint railway signs, such as 'Whistle'.

CHARLES CARSON

Appendix VI

My first solo flight in Tiger Moth G-AIRR took place on the seventeenth of February 1960 after eight hours and forty-five minutes instruction.

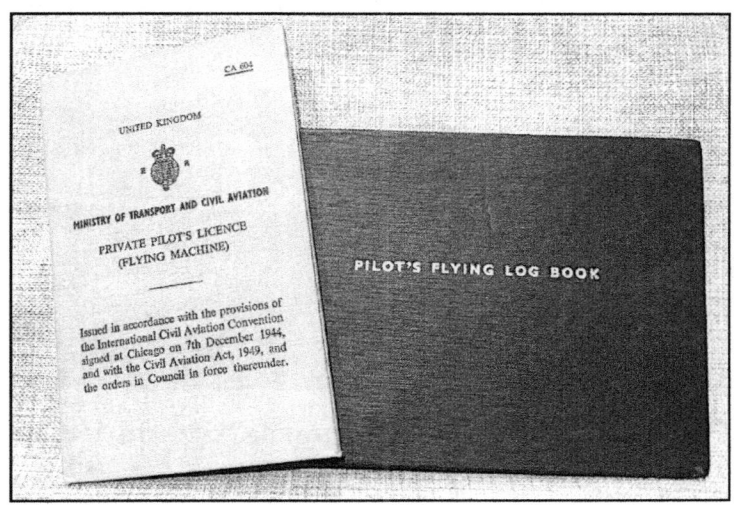

Plate 79. Pilot's Licence & Log Book

My Private Pilot's Licence was awarded after the requisite forty hours flying time and tests had been completed. Tubby Dash was my instructor, he taught me well.

 https://youtu.be/-i8tUgNlz_M

It is hard to believe now that I had the courage in 1963 to fly a bright yellow G-AOUR Tiger Moth from Newtownards Airfield to Belfast Lough and practice loops, stall recoveries and steep turns. I was twenty-five years old and engaged to be married to Ann. See Plate 80.

Plate 80. G-AOUR Newtownards Airfield 1963

Ann and I were married on sixth of August 1964. I never flew again. Other adventures called.

---

[i] C.C Dash (Tubby). DASH, Cyril Charles Henry, WO - 24 EFTS - RAFVR

Printed in Great Britain
by Amazon